I0471149

U.S. Department
of Transportation

**National Highway
Traffic Safety
Administration**

www.nhtsa.gov

DOT HS 811 500

July 2011

Run-Off-Road Crashes:
An On-Scene Perspective

1. Report No. DOT HS 811 500	2. Government Accession No.		3. Recipient's Catalog No.
4. Title and Subtitle **Run-Off-Road Crashes: An On-Scene Perspective**		5. Report Date July 2011	
		6. Performing Organization Code NVS-421	
7. Author(s) **Cejun Liu, Ph.D., and Tony Jianqiang Ye** Statisticians, Bowhead Systems Management Inc., contractors working with Mathematical Analysis Division, National Center for Statistics and Analysis, National Highway Traffic Safety Administration		8. Performing Organization Report No.	
9. Performing Organization Name and Address Mathematical Analysis Division, National Center for Statistics and Analysis National Highway Traffic Safety Administration 1200 New Jersey Avenue SE., Washington, DC 20590		10. Work Unit No. (TRAIS)n code	
		11. Contract of Grant No.	
12. Sponsoring Agency Name and Address Mathematical Analysis Division, National Center for Statistics and Analysis National Highway Traffic Safety Administration 1200 New Jersey Avenue SE., Washington, DC 20590		13. Type of Report and Period Covered NHTSA Technical Report	
		14. Sponsoring Agency Code	
15.Supplementary Notes			

16. Abstract

Run-off-road (ROR) crashes, which usually involve only a single vehicle, contribute to a large portion of fatalities and serious injuries to motor vehicle occupants. In this study, the National Motor Vehicle Crash Causation Survey (NMVCCS) data collected at crash scenes between 2005 and 2007 is used to identify the ROR critical pre-crash event, assess the critical reason for the ROR critical event, and examine associated factors present in the pre-crash phase of the ROR crash. The effect of antilock brake system (ABS) and electronic stability control (ESC) on ROR crashes is also evaluated.

The results show that over 95 percent of the critical reasons for single-vehicle ROR crashes were driver-related. The most frequently occurring category of critical reasons attributed to drivers was driver performance errors (27.7%) such as "overcompensation" and "poor directional control," followed by driver decision errors (25.4%) such as "too fast for curve" and "too fast for conditions," critical non-performance errors (22.5%) such as "sleeping" and "heart attack/other physical impairment," and recognition errors (19.8%) such as "internal distractions" and "external distractions." With the presence of alcohol in the driver, as high as 46.9 percent of driver-related critical reasons for single-vehicle ROR crashes were driver performance errors. The logistic regression analysis shows that the most influential factors in the occurrence of single-vehicle ROR crashes were the factors "driver inattention," "driver was fatigued," and "driver was in a hurry."

In the NMVCCS crashes, for the vehicles equipped with both ABS and ESC, 7.5 percent ran off the road, while for the vehicles equipped with neither ABS nor ESC, 14.6 percent ran off the road. The odds of being involved in ROR crashes for the vehicles equipped with neither ABS nor ESC were 2.1 times greater than the odds for the vehicles equipped with both ABS and ESC. The combined effect of ABS and ESC systems on reducing the ROR crashes is significant, which is consistent with prior evaluation of the long-term effect of ABS and ESC based on the FARS and GES data. This study is NHTSA's first effort in evaluating the effectiveness of crash avoidance technologies with the NMVCCS data.

17. Key Words Run-off-road, Pre-crash event, Single-vehicle crashes, NMVCCS, Critical reason, Crash-associated factor, LTCCS, ABS, ESC		18. Distribution Statement Document is available to the public from the National Technical Information Service www.ntis.gov	
19. Security Classif. (of this report) Unclassified	20. Security Classif. (of this page) Unclassified	21. No of Pages 36	22. Price

i

Table of Contents

NHTSA's National Center for Statistics and Analysis   1200 New Jersey Avenue SE., Washington, DC 20590

List of Figures

NHTSA's National Center for Statistics and Analysis   1200 New Jersey Avenue SE., Washington, DC 20590

List of Tables

NHTSA's National Center for Statistics and Analysis   1200 New Jersey Avenue SE., Washington, DC 20590

Executive Summary

A run-off-road (ROR) crash occurs when a vehicle in transit leaves the road and collides with a tree, a pole, other natural or artificial objects, or overturns on non-traversable terrain. This type of crash usually involves only a single vehicle. ROR crashes account for a large portion of serious injuries and fatalities among all traffic crashes.

In this study, data from the National Motor Vehicle Crash Causation Survey (NMVCCS) for fatal and nonfatal crashes involving passenger vehicles (passenger cars, vans, pickup trucks, and sport utility vehicles) are used. The 5,470 unweighted crashes from the NMVCCS that were assigned sampling weights represent 2,188,979 crashes at the national level. A single-vehicle ROR crash is defined with the variable "critical pre-crash event" as "the vehicle ran off the left or right edge of the road." The critical pre-crash event identifies the event that made the crash imminent (i.e., something occurred that made the crash immediate and unavoidable). For the purpose of comparison, all the other critical events (excluding "unknown") are referred to as "Other." The crashes with the "Other" critical pre-crash events are generally the on-road crashes in which the vehicle remained on the road after the crash.

The ROR critical pre-crash event is identified and the critical reason underlying the critical pre-crash event as well as the crash-associated factors is examined. An assessment of the critical reasons for the large-truck single-vehicle ROR crashes based on the Large-Truck Crash Causation Study (LTCCS) is presented in this study for the purpose of comparison. The effect of antilock brake system (ABS) and electronic stability control (ESC) on ROR crashes is also evaluated.

The following are some of the findings from the descriptive, contingency, and logistic regression analyses conducted in this study using the NMVCCS data:

- ROR crashes accounted for 64.4 percent of all single-vehicle crashes.

- Among the ROR single-vehicle crashes, 95.1 percent of the critical reasons were driver-related, while among the "Other" single-vehicle crashes, 84.1 percent were driver-related.

- The most frequently occurring category of the critical reasons attributed to drivers in single-vehicle ROR crashes was driver performance errors (27.7%), followed by driver decision errors (25.4%), critical nonperformance errors (22.5%), and driver recognition errors (19.8%). In contrast, driver decision errors (59.7%) and performance errors (26.3%) were the top two categories of the driver-related critical reasons for the "Other" single-vehicle crashes.

- With the presence of alcohol in the driver, as high as 46.9 percent of driver-related critical reasons for the single-vehicle ROR crashes were driver performance errors such as "overcompensation" and "poor directional control."

- For passenger vehicles (based on the NMVCCS), the dominant critical reasons for single-vehicle ROR crashes were "internal distraction" (14.3%), "overcompensation" (13.6%), "poor directional control" (12%), "too fast for curve" (10.5%), and "sleeping/actually asleep" (9.8%). All these critical reasons were driver-related.

v

- For large trucks (based on the LTCCS), "sleeping/actually asleep" (33.1%) and "heart attack/other physical impairment" (14.9%) were the most frequently assigned critical reasons for the single-vehicle ROR crashes.

- Factors significantly associated with the occurrence of ROR crashes include driver inattention, driver fatigue status, roadway surface conditions, driver alcohol presence, driver's familiarity with the roadway, driver's pre-existing physical or mental health conditions, driver's gender, driver's work-related stress or pressure, and if the driver was in a hurry.

- Logistic regression is used to assess the relative influence of the crash-associated factors. It shows that the most influential factors in the occurrence of single-vehicle ROR crashes were the factors "driver inattention" (odds ratio=3.66), "driver was fatigued" (odds ratio=3.48), and "driver was in a hurry" (odds ratio=3.20).

- In the NMVCCS crashes, for the vehicles equipped with both ABS and ESC, 7.5 percent ran off the road, while for the vehicles equipped with neither ABS nor ESC, 14.6 percent ran off the road. The odds ratio of 2.1, derived from the two percentages, indicates that the odds of being involved in ROR crashes for the vehicles equipped with neither ABS nor ESC were 2.1 times greater than the odds for the vehicles equipped with both ABS and ESC. This is statistically significant at the 90 percent confidence level. Thus, one could infer that the combined effect of ABS and ESC systems on reducing the ROR crashes is significant.

NHTSA's National Center for Statistics and Analysis   1200 New Jersey Avenue SE., Washington, DC 20590

1. Introduction

A vehicle in transport sometimes leaves the road and hits one or more natural or artificial objects. This event usually involves only a single vehicle and is referred to as a run-off-road (ROR) crash. ROR crashes contribute to a large portion of fatalities and injuries to the vehicle occupants in motor vehicle traffic crashes. According to the 2008 NHTSA data from the Fatality Analysis Reporting System (FARS), 60 percent of all fatal crashes were single-vehicle crashes and 71 percent of these fatal single-vehicle crashes were ROR crashes.[1]

The studies[2,3,4,5] conducted in the past have shed some light on a variety of factors that are closely associated with the occurrence of ROR crashes. For instance, a 2009 NHTSA report[5] found that factors such as curved road segments, rural roads, high-speed-limit roadways, adverse weather, and alcohol use by drivers were associated with a high risk of fatal single-vehicle ROR crashes.

The scope of this study is not limited to fatal crashes, but covers both fatal and nonfatal crashes from the NMVCCS.[6] Unlike the previous studies that mainly explored environment-related factors, this study thoroughly investigates driver-, vehicle-, and environment-related factors, with a focus on factors that are related to driver's physical and mental conditions, as well as driver's activities prior to the crash. It could only be made possible by the NMVCCS data in which driver-, vehicle-, weather-, and roadway-related on-scene information was collected immediately after the crash occurrence so as to avoid loss or distortion of information due to lapse of time. The NMVCCS was conducted by NHTSA's National Center for Statistics and Analysis and investigated a total of 6,949 crashes during the period January 2005 to December 2007.

The goal of this study is to identify the ROR critical pre-crash event, assess the critical reason for this critical event, and examine the associated factors present in the pre-crash phase of the single-vehicle ROR crash. Additionally, the effect of the antilock brake system (ABS) and the electronic stability control (ESC) system on ROR crashes is evaluated.

The target population for this study consists of passenger vehicles involved in the ROR crashes. However, for the purpose of comparison, an assessment of the critical reasons for the large-truck single-vehicle ROR crashes, based on the LTCCS, is also presented in the study. The LTCCS, conducted by Federal Motor Carrier Safety Administration and NHTSA, investigated a total of 1,070 crashes during the period April 2001 to December 2003. Each of these crashes involved at least one large truck and resulted in at least one fatality or one evident injury.

The outline of this report is as follows. Section 2 briefly describes the data and the methodology used in this report. Section 3 presents statistics of the ROR critical pre-crash events and critical reasons underlying these critical events in single-vehicle crashes. In Section 4, a descriptive (univariate) analysis is conducted to study the crash-associated factors. The amount of risk associated with each factor (odds ratio) in the occurrence of ROR crashes is assessed using the logistic regression. The effect of ABS and ESC on ROR crashes is evaluated in Section 5. The summary and conclusions are presented in Section 6.

NHTSA's National Center for Statistics and Analysis   1200 New Jersey Avenue SE., Washington, DC 20590

2. Data and Methodology

2.1 The NMVCCS data

The data used in this study comes from the NMVCCS. Of the 6,949 crashes investigated in the NMVCCS, 5,470 were assigned sampling weights to provide national estimates, while the remaining 1,479 unweighted crashes can be used for clinical studies. The weighted crashes, investigated during the period July 2005 and December 2007, represent 2,188,970 crashes at the national level involving 4,031,075 vehicles and 3,944,621 drivers.

Understanding the events leading up to a crash is essential for crash prevention. The NMVCCS collected information from the chain of events preceding the "first harmful event" (i.e., the first event during the crash occurrence that caused injury or property damage). This crash causal chain is characterized by four elements: movement prior to critical crash envelope, critical pre-crash event, critical reason for the critical pre-crash event, and other crash-associated factors. Among these, the critical pre-crash event identifies the event that made the crash imminent (i.e., something occurred that made the crash inevitable), and is coded for each of the first three in-transport vehicles referred to as case vehicles. The critical reason is the immediate reason for the critical event and is often the last failure in the causal chain (i.e., closest in time to the critical pre-crash event). The critical reason, assigned to one of the case vehicles in a crash, can be attributed to the driver, vehicle, weather, or roadway condition.[6] However, it is important to keep in mind that the critical event, the critical reason, or the associated factors should not be interpreted as the cause of the crash or an assignment of the fault to the driver, vehicle, or environment.

Based on the above perspective, the NMVCCS investigated crashes involving light passenger vehicles. In an NMVCCS crash, at least one of the first three vehicles involved in the crash was towed or would be towed due to damage. The NMVCCS data includes more than 600 variables or factors related to drivers, vehicles, and environment.[6] In order to facilitate the timely collection of on-scene crash data, NMVCCS researchers attempt to arrive at crash scenes before they are cleared and begin collecting data through scene inspection, photographs, interviews of drivers and witnesses, and a limited vehicle inspection.

In this study, a single-vehicle ROR crash refers to a crash in which the critical pre-crash event is "the vehicle ran off the left or right edge of the road." For the purpose of comparison, all the other critical pre-crash events (excluding "unknown") are referred to as "Other." These non-ROR pre-crash events include the events characterized by "vehicle loss of control due to blow out/flat tire, poor road condition and other cause," "vehicle turning at or passing through intersection," "pedestrian, pedal-cyclist, or other non-motorist in or approaching roadway," and "animal in or approaching roadway." The crashes with "Other" critical pre-crash events are generally on-road crashes in which after the crash the vehicles remained on the road.

Except for the comparison between the NMVCCS and LTCCS data on critical reasons in Section 3, all other analyses in this report pertain to ROR crashes that involve passenger vehicles, including passenger cars, vans, pickup trucks, and sport utility vehicles (SUV) with gross vehicle weight ratings under 10,000 pounds.

NHTSA's National Center for Statistics and Analysis   1200 New Jersey Avenue SE., Washington, DC 20590

For single-vehicle crashes, since only one pre-crash critical event is coded for the vehicle, the terms crash, vehicle, driver, and pre-crash event are used interchangeably.

ROR Crash - An NMVCCS Case Example

The following case example illustrates a single-vehicle ROR crash and the manner in which it was coded in the NMVCCS.

Case description: A crash involving a 2004 Subaru Forester (a compact SUV) occurred on the late weekday afternoon on a dry roadway with a posted speed limit of 40 mph (64 km/h). The Subaru driver was an 82-year-old male. He had some pre-existing physical or mental health condition and reported taking drugs/medications in the past 24 hours. The driver tried to avoid a non-contact truck approached from the opposite direction by steering right and the vehicle ran off the edge of the road on the right side. The vehicle was equipped with ABS but not with ESC.

NMVCCS coding

- Critical pre-crash event – "this vehicle ran off the edge of the road on the right side"
- Critical reason for the critical event – "poor directional control (e.g., failing to control vehicle with skill ordinarily expected)"
- Crash-associated factors: pre-existing physical or mental health conditions; taking drugs/medications in the 24 hours; attempted an avoidance maneuver by steering right; age; gender …

2.2 The Methodology

In this study, statistics for the ROR and "Other" critical pre-crash events, and the frequency distributions of the critical reasons for these events are presented. Descriptive (univariate) analysis is conducted to study several crash-associated factors. The Wald chi-square test is used to assess whether the differences in percentages between the dichotomies of each associated factor are statistically significant at the 90 percent confidence level. The impact of these factors is also assessed by logistic regression in which the "Other" crash events essentially form one element of the binary outcome (ROR versus "Other"). This procedure helps to assess their relative influence as well as estimate the amount of risk each factor carried (odds ratio) in the occurrence of ROR crashes.

Frequency (percentage) tables in the following sections are all based on the weighted data. Due to the complex nature of the NMVCCS sample design, the SURVEYFREQ and SURVEYLO-GISTIC procedures of SAS Version 9.1 are used.

NHTSA's National Center for Statistics and Analysis   1200 New Jersey Avenue SE., Washington, DC 20590

3. Critical Reasons for the Single-Vehicle ROR Crash Events

A critical reason is the immediate reason for a pre-crash event and is often the last failure in the causal chain (i.e., closest in time to the critical pre-crash event). The critical reason is normally coded to only one vehicle in each crash. It can be assigned to driver (e.g., performance error, decision error, recognition error, critical non-performance error, or unknown driver error), vehicle (failure), or environment (roadway or weather).

Statistics for critical reasons of the single-vehicle ROR and "Other" critical pre-crash events are presented in Table 1. ROR crashes accounted for 64.4 percent (434,412) of the estimated 674,002 single-vehicle crashes. Among all critical reasons coded for the single-vehicle ROR crashes, those attributed to drivers predominated (95.1%). With unknown reasons excluded, 97.8 percent are attributed to drivers. This provides evidence of the importance of driver-related factors in traffic crashes.

These predominant driver-related critical reasons are discussed in detail in Section 3.1. The breakdown of critical reasons attributed to vehicles, and environment is presented in Section 3.2 and 3.3, respectively.

Table 1: Critical Reasons Coded for the Single-Vehicle ROR and "Other" Crash Events

Critical Reasons Attributed to	ROR		"Other"	
	Weighted Frequency	Weighted Percent	Weighted Frequency	Weighted Percent
Driver	413,070	95.1%	201,408	84.1%
Vehicle	4,456	1.0%	20,631	8.6%
Environment (roadway and weather conditions)	4,950	1.1%	16,385	6.8%
Unknown reason for the critical event	11,937	2.7%	1,087	0.5%
Critical reason not coded to the vehicle	0	0.0%	80	0.0%
Total	434,412	Col. 100% / Row 64.4%	239,590	Col. 100% / Row 35.6%

Note: Estimates may not add up to totals due to independent rounding.
Data source: NMVCCS (2005–2007)

3.1 Critical Reasons Attributed to Drivers

About 95.1 percent of the critical reasons were attributed to drivers in single-vehicle ROR crashes (Table 1). Table 2 shows the weighted frequencies of the five categories of driver-related critical reasons, namely critical non-performance errors, recognition errors, decision errors, performance errors, and unknown driver errors.

Of the 413,070 single-vehicle ROR crashes in which critical reason was attributed to drivers, the most frequently occurring category was driver performance errors (27.7%), followed by driver decision errors (25.4%), critical non-performance errors (22.5%), and recognition errors (19.8%). In contrast, driver decision errors (59.7%) and performance errors (26.3%) were the two most

4

frequently occurring categories of driver-related critical reasons for the "Other" single-vehicle crashes.

Among driver performance errors, "overcompensation" (14.3%) and "poor directional control" (12.6%) were the top two critical reasons for single-vehicle ROR crashes.

Among driver decision errors, the most frequently occurring critical reasons for single-vehicle ROR crashes were "too fast for curves" (11%), "too fast for conditions" (6.8%), and "incorrect evasion" (3.3%).

Among critical non-performance errors, the most frequently occurring critical reasons for single-vehicle ROR crashes were "sleeping/actually asleep" (10.3%) and "heart attack or other physical impairment" (7.1%).

Among driver recognition errors (the driver failed to correctly recognize the pre-crash situation), "internal distraction" (15%) and "external distraction" (2.7%) were the major critical reasons for single-vehicle ROR crashes.

NHTSA's National Center for Statistics and Analysis   1200 New Jersey Avenue SE., Washington, DC 20590

Table 2: Critical Reasons for the Single-Vehicle ROR and "Other" Crash Events Attributed to Drivers

	Critical Reasons	ROR		"Other"	
		Weighted Frequency	Weighted Percent	Weighted Frequency	Weighted Percent
Critical Non-Performance Errors	Sleeping/actually asleep	42,586	**10.3%**	886	0.4%
	Heart attack/other physical impairment	29,226	7.1%	1,646	0.8%
	Other/unknown critical nonperformance	20,961	5.1%	311	0.2%
	Subtotal	92,773	22.5%	2,843	1.4%
Recognition Errors	Internal distraction	62,048	**15.0%**	10,561	5.2%
	External distraction	11,324	2.7%	591	0.3%
	Inattention	5,644	1.4%	2,262	1.1%
	Inadequate surveillance	1,651	0.4%	8,303	4.1%
	Other/unknown recognition error	1,313	0.3%	545	0.3%
	Subtotal	81,980	19.8%	22,262	11.0%
Decision Errors	Too fast for curve	45,429	**11.0%**	39,813	19.8%
	Too fast for conditions	27,983	6.8%	55,092	27.4%
	Incorrect evasion	13,529	3.3%	8,626	4.3%
	Aggressive driving	6,894	1.7%	4,813	2.4%
	Too fast to be able to respond	5,314	1.3%	5,819	2.9%
	Inadequate evasion	2,173	0.5%	2,450	1.2%
	Other/unknown decision error	1,432	0.4%	1,644	0.8%
	Illegal maneuver	203	0.1%	1,074	0.5%
	Misjudgment of gap	915	0.2%	42	0.0%
	Following too closely	334	0.1%	700	0.4%
	Subtotal	104,206	25.4%	120,073	59.7%
Performance Errors	Overcompensation	59,155	**14.3%**	31,410	15.6%
	Poor directional control	51,991	**12.6%**	19,004	9.4%
	Other/unknown performance error	2,088	0.5%	1,193	0.6%
	Panic/freezing	1,149	0.3%	1,346	0.7%
	Subtotal	114,383	27.7%	52,953	26.3%
Other/Unknown Driver Errors		19,726	4.8%	3,276	1.6%
Total		413,070	100%	201,408	100%

Note: Estimates may not add up to totals due to independent rounding.
Data source: NMVCCS (2005 – 2007)

The dominant individual critical reasons (in descending order) for single-vehicle ROR crashes attributed to drivers were "internal distraction," "overcompensation," "poor directional control," "too fast for curve," and "sleeping/actually asleep" as shown by the highlighted weighted percentages in Table 2. It is consistent with the findings from the 2009 NHTSA report on fatal single-vehicle ROR crashes.[5]

Driver Alcohol Presence

According to the 2009 NHTSA report and some other studies[2][5] in the literature, drivers with alcohol use are more likely to be involved in ROR crashes as compared to sober drivers. Driver alcohol presence as an associated factor in ROR crashes will be discussed in Section 4. This section investigates the critical reasons for the single-vehicle ROR crash events with and without the presence of alcohol for the driver.

NHTSA's National Center for Statistics and Analysis   1200 New Jersey Avenue SE., Washington, DC 20590

Table 3 shows that the presence of alcohol in the driver affected the driver's performance in single-vehicle ROR crashes. With the presence of alcohol in the driver, 23.4 percent (as compared to 12.6% without alcohol) and 21.7 percent (as compared to 11.2% without alcohol) of the driver-related critical reasons were "overcompensation" and "poor directional control" in single-vehicle ROR crashes. Category-wise, with the presence of alcohol, 46.9 percent (as compared to 24.4% without alcohol) of driver-related critical reasons for single-vehicle ROR crashes were driver performance errors.

Table 3: Critical Reasons for the Single-Vehicle ROR Crash Events Attributed to Drivers With Versus Without the Presence of Alcohol in the Driver

	Critical Reasons	Alcohol Present		Alcohol Not Present	
		Weighted Frequency	Weighted Percent	Weighted Frequency	Weighted Percent
Critical Non-Performance Errors	Sleeping/actually asleep	3,220	4.2%	37,795	12.2%
	Heart attack/other physical impairment	794	1.0%	27,650	9.0%
	Other/unknown critical nonperformance	8,377	11.0%	11,890	3.9%
	Subtotal	12,391	16.2%	77,335	25.1%
Recognition Errors	Internal distraction	10,578	13.9%	44,979	14.6%
	External distraction	961	1.3%	10,070	3.3%
	Inattention	0	0%	5,082	1.7%
	Inadequate surveillance	0	0%	1,651	0.5%
	Other/unknown recognition error	1,038	1.4%	275	0.1%
	Subtotal	12,577	16.6%	62,057	20.2%
Decision Errors	Too fast for curve	6,447	8.5%	36,331	11.8%
	Too fast for conditions	5,436	7.1%	22,314	7.2%
	Incorrect evasion	0	0%	13,529	4.4%
	Aggressive driving	767	1.0%	6,127	2.0%
	Too fast to be able to respond	718	0.9%	4,268	1.4%
	Inadequate evasion	0	0%	2,009	0.7%
	Other/unknown decision error	0	0%	1,432	0.5%
	Illegal maneuver	0	0%	203	0.1%
	Misjudgment of gap	0	0%	915	0.3%
	Following too closely	0	0%	334	0.1%
	Subtotal	13,368	17.5%	87,462	28.5%
Performance Errors	Overcompensation	17,864	**23.4%**	39,057	12.6%
	Poor directional control	16,564	**21.7%**	34,657	11.2%
	Other/unknown performance error	1,391	1.8%	698	0.2%
	Panic/freezing	0	0%	1,149	0.4%
	Subtotal	35,819	**46.9%**	75,561	24.4%
Other/Unknown Driver Errors		2,129	2.8%	6,518	2.1%
Total		76,283	100%	308,932	100%

Note: Estimates may not add up to totals due to independent rounding.
Data source: NMVCCS (2005 – 2007)

NHTSA's National Center for Statistics and Analysis   1200 New Jersey Avenue SE., Washington, DC 20590

3.2 Critical Reasons Attributed to Vehicles

Only 1.0 percent of the critical reasons were attributed to vehicles in single-vehicle ROR crashes (Table 1). Table 4 presents statistics of the ROR and "Other" crash events in which the critical reason was attributed to vehicles in single-vehicle crashes. In ROR crashes, the most frequently occurring critical reason attributed to vehicles was "brakes failed/degraded" (32.7%), followed by "tires failed or degradation/wheel failed" (25.6%), "steering/suspension/transmission/engine failed" (19.1%), and "other vehicle failure/deficiency" (18.8%).

In contrast, in the "Other" single-vehicle crashes, the most frequently occurring vehicle-related critical reason was "tires failed or degradation/wheel failed" (71.7%).

Table 4: Critical Reasons for the Single-Vehicle ROR and "Other" Crash Events Attributed to Vehicles

Critical Reasons	ROR		"Other"	
	Weighted Frequency	Weighted Percent	Weighted Frequency	Weighted Percent
Tires failed or degradation/wheel failed	1,142	25.6%	14,790	71.7%
Steering/suspension/transmission/engine failed	850	19.1%	3,272	15.8%
Brakes failed/degraded	1,457	32.7%	2,155	10.5%
Other vehicle failure/deficiency	839	18.8%	413	2.0%
Unknown vehicle failure	167	3.8%	0	0.0%
Total	4,456	100%	20,631	100%

Note: Estimates may not add up to totals due to independent rounding.
Data source: NMVCCS (2005–2007)

3.3 Critical Reasons Attributed to Environment

Only 1.1 percent of the critical reasons were attributed to environment in single-vehicle ROR crashes (Table 1). Table 5 presents statistics of the ROR and "Other" crash events in which the critical reason was attributed to environment (roadway and weather conditions) in single-vehicle crashes. Among the ROR crashes in which the critical reason was attributed to environment, about 96.5 percent were related to roadway conditions (slick roads, 64.3%; other highway-related conditions, 32.2%) while only 3.5 percent to the weather (predominantly rain or snow.)

In contrast, among the "Other" single-vehicle crashes in which the critical reason was attributed to environment, 84.2 percent were related to roadway conditions and 15.8 percent to weather conditions.

8

Table 5: Critical Reasons for the Single-Vehicle ROR and "Other" Crash Events Attributed to Environment

	Critical Reasons	ROR		"Other"	
		Weighted Frequency	Weighted Percent	Weighted Frequency	Weighted Percent
Roadway	Slick roads (ice, loose debris, etc.)	3,183	64.3%	11,942	72.9%
	Other highway-related (sign/signal/road design/view obstructions, etc) conditions	1,592	32.2%	1,843	11.3%
	Subtotal	4,775	96.5%	13,785	84.2%
Weather	Rain/snow	174	3.5%	1,035	6.3%
	Other weather-related (fog/glare/wind, etc) condition	0	0.0%	1,565	9.5%
	Subtotal	174	3.5%	2,600	15.8%
Total		4,950	100%	16,385	100%

Note: Estimates may not add up to totals due to independent rounding.
Data source: NMVCCS (2005 – 2007)

3.4 Critical Reasons for ROR Crashes Involving Large Trucks, Based on the LTCCS

The focus of this study is on passenger vehicles involved in ROR crashes investigated in the NMVCCS. However, for comparison purposes, the critical reasons for single-vehicle ROR crashes involving large trucks (with a gross weight rating of over 10,000 pounds) are also studied. The LTCCS[7] data is used for this purpose. Like the NMVCCS, the LTCCS collected the driver-, vehicle-, and environment-related on-scene information.[8]

The LTCCS data is based on 1,070 crashes during the period from April 2001 to December 2003. Each crash involved at least one large truck with a gross vehicle weight rating of over 10,000 pounds, and resulted in at least one fatality or one evident injury. The LTCCS collected data on approximately 1,000 variables for each crash.[7, 9]

Statistics of critical reasons for the large-truck single-vehicle ROR and "Other" crash events are presented in Table 6. Based on the LTCCS data, ROR crashes accounted for only 37.2 percent of all single-vehicle crashes involving large trucks. In contrast, based on the NMVCCS data, ROR crashes accounted for 64.4 percent of all single-vehicle crashes involving passenger vehicles.

Table 6 also shows that among the critical reasons for the large-truck single-vehicle ROR crashes, the most frequently occurring critical reason is "sleeping/actually asleep" (33.1%), followed by "heart attack/other physical impairment incurred to the driver" (14.9%). In contrast, among the critical reasons for the "Other" large-truck single-vehicle crashes, 32.3 percent were "too fast for curve/turn" and 9.3 percent "the cargo shifted."

A comparison of the most frequently occurring critical reasons underlying the single-vehicle ROR crash events between passenger vehicles (based on the NMVCCS) and large trucks (based on the LTCCS) is presented in Table 7. It shows that "internal distraction" (14.3%), "overcompensation" (13.6%), "poor directional control" (12%), "too fast for curve" (10.5%), and "sleeping/actually asleep" (9.8%) were the major critical reasons for passenger vehicle single-vehicle

NHTSA's National Center for Statistics and Analysis   1200 New Jersey Avenue SE., Washington, DC 20590

ROR crashes. In the case of large-truck single-vehicle ROR crashes, "sleeping/actually asleep" (33.1%) and "heart attack/other physical impairment" (14.9%) were the most frequently assigned critical reasons. These high rates are probably due to the fact that drivers of large trucks usually drive long-distances on highways.[10]

Table 6: Critical Reasons for the Large-Truck Single-Vehicle ROR and "Other" Crash Events

Critical Reasons	ROR		Other	
	Weighted Frequency	Weighted Percent	Weighted Frequency	Weighted Percent
Driver-Related				
Too Fast for curve/turn	782	5.5%	7,719	**32.3%**
Sleeping /actually asleep	4,696	**33.1%**	180	0.8%
Too fast for conditions to be able to respond...	964	6.8%	1,494	6.2%
Inattention (i.e., daydreaming)	963	6.8%	1,290	5.4%
Heart attack or other physical impairment of the ability...	2,114	**14.9%**	127	0.5%
Overcompensation	479	3.4%	1,117	4.7%
Poor directional control e.g., failing to control vehicle...	570	4.0%	943	3.9%
Inadequate surveillance (e.g., failed to look, looked but...)	0	0.0%	1,376	5.8%
Internal distraction	125	0.9%	1,160	4.8%
Aggressive driving behavior	0	0.0%	814	3.4%
Other decision error	295	2.1%	190	0.8%
External distraction	272	1.9%	69	0.3%
Illegal maneuver	0	0.0%	157	0.7%
Following too closely to respond to unexpected actions	144	1.0%	0	0.0%
Other critical non-performance	138	1.0%	0	0.0%
Misjudgment of gap or other's speed	0	0.0%	62	0.3%
Type of driver error unknown	1,248	8.8%	69	0.3%
Unknown recognition error	396	2.8%	659	2.7%
Unknown critical non-performance	489	3.4%	0	0.0%
Vehicle-Related				
Cargo shifted	280	2.0%	2,223	**9.3%**
Tires/wheels failed	0	0.0%	387	1.6%
Brakes failed	94	0.7%	230	1.0%
Steering failed	0	0.0%	69	0.3%
Degraded braking capability	0	0.0%	758	3.2%
Suspension failed	0	0.0%	791	3.3%
Environment (roadway and weather conditions)				
Road design – other	0	0.0%	356	1.5%
Slick roads (low friction road surface due to ice...)	125	0.9%	115	0.5%
Wind gust	0	0.0%	127	0.5%
Road design - roadway geometry (e.g., ramp curvature)	2	0.0%	0	0.0%
Unknown reason for critical event	23	0.2%	0	0.0%
Critical event not coded to this vehicle	0	0.0%	1,447	6.1%
Total	14,198	Col. 100% / Row 37.2%		23,928	Col. 100% / Row 62.8%

Note: Estimates may not add up totals due to independent rounding.
Data source: LTCCS (2001 – 2003)

NHTSA's National Center for Statistics and Analysis   1200 New Jersey Avenue SE., Washington, DC 20590

Table 7: A Comparison of Major Critical Reasons for the Passenger Vehicle and Large-Truck Single-Vehicle ROR Crash Events

Critical Reasons	Passenger Vehicle (NMVCCS)	Large Truck (LTCCS)
	Weighted Percent	Weighted Percent
Sleep, that is, actually asleep	9.8%	**33.1%**
Heart attack/other physical impairment	6.7%	**14.9%**
Internal distraction	**14.3%**	0.9%
Overcompensation	**13.6%**	3.4%
Poor directional control	**12.0%**	4.0%
Too Fast for curve	**10.5%**	5.5%
Inattention	1.3%	**6.8%**
External distraction	**2.6%**	1.9%

Note: The frequencies for Passenger Vehicle from the NMVCCS reported in this table are slightly different from those in Table 2. This difference is because the frequencies in this table are percentages of **all** critical reasons (**combined** driver, vehicle, and environment related critical reasons) while the frequencies in Table 2 are percentages of driver-related critical reasons.
Data source: NMVCCS (2005–2007)
                LTCCS (2001–2003)

11

4. Crash-Associated Factors in Single-Vehicle ROR Crashes

In addition to the critical reasons, the NMVCCS data provides information about the driver-, vehicle-, and environment-related associated factors that were present in the pre-crash phase. This section focuses on the driver-related factors that pertain to driver's physical and mental conditions. Of particular interest are: driver inattention, driver alcohol presence, driver fatigue status, driver's gender, driver's work-related stress or pressure, driver's pre-existing physical and/or mental health conditions, driver's familiarity with the roadway, and whether the driver was in a hurry. One of the environment-related factors, roadway surface conditions, is also examined in this study.

For other factors such as driver drug use and cell phone use, the statistics show that these factors are associated with high risk of single-vehicle ROR crashes, though the association in these cases is not statistically significant (data not shown in this report).

Descriptive (univariate) analysis is conducted to study the crash-associated factors. The relative influence as well as the amount of risks each factor carried (odds ratio) in the occurrence of such crashes is assessed by logistic regression modeling.

4.1 Descriptive Analysis

The factors examined below are driver inattention, driver alcohol presence, driver fatigue status, driver's gender, driver's work-related stress or pressure, driver's pre-existing physical and/or mental health conditions, driver's familiarity with the roadway, roadway surface conditions, and if the driver was in a hurry. To study the effect of each of these factors, the single vehicle crashes were categorized into two groups: the crashes in which the factor was present and those in which it was not present. The Wald chi-square test is used to assess if the differences in percentages between the two groups are statistically significant. For each of the examined factors, since it might not be coded in all NMVCCS crashes (i.e., missing or unknown values), the total crashes for the individual factor in Figures 1 through 9 may not add up to the totals of ROR and "Other" crashes as shown in Table 2.

Inattention

In the NMVCCS, a driver was assessed inattentive if he or she was pre-occupied with concerns or the nature of these concerns such as personal problems, family problems, financial problems, preceding arguments, and future events (e.g., vacation and wedding). Figure 1 presents the frequencies of single-vehicle ROR and "Other" crashes by driver inattention.

Among single-vehicle crashes in which drivers were inattentive, 85.4 percent were ROR crashes, while in single-vehicle crashes that did not involve driver inattention, only 57.1 percent were ROR crashes. The difference between these two percentages is statistically significant at the 90 percent confidence level (χ^2=4.23, p-value=0.0622). This indicates that inattention while driving is significantly associated with ROR crashes.

NHTSA's National Center for Statistics and Analysis   1200 New Jersey Avenue SE., Washington, DC 20590

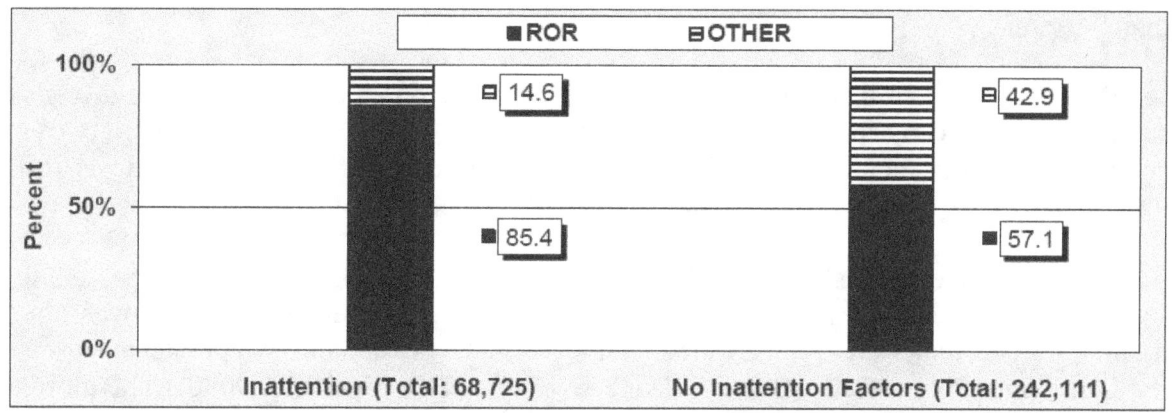

Figure 1: Single-Vehicle Crashes by Driver Inattention

(Crashes With Missing and Unknown Values for the Factor Not Included)

Driver Alcohol Presence

In the NMVCCS, the variable "police reported alcohol presence" records the presence of alcohol for the driver as reported by police in the police accident report (PAR). Figure 2 presents the frequencies of single-vehicle ROR and "Other" crashes by driver alcohol presence.

Among single-vehicle crashes with the presence of alcohol for the driver, 83.6 percent were ROR crashes, while in single-vehicle crashes without driver alcohol presence, only 60.9 percent were ROR crashes. The difference between these two percentages is statistically significant at the 90 percent confidence level (χ^2=19.26, p-value=0.0009). This indicates that driver alcohol presence is significantly associated with ROR crashes.

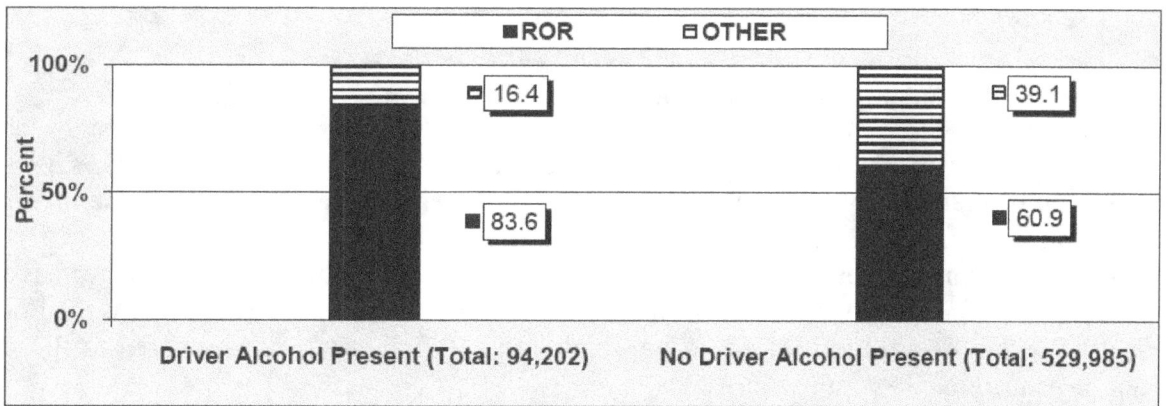

Figure 2: Single-Vehicle Crashes by Driver Alcohol Presence

(Crashes With Missing and Unknown Values for the Factor Not Included)

13

Fatigue

In the NMVCCS, the fact that a driver was fatigued is based on an evaluation of the driver's current and preceding sleep schedules, current and preceding work schedules, and a variety of other fatigue-related factors including recreational and non-work activities. Figure 3 shows the frequencies of single-vehicle ROR and "Other" crashes by driver fatigue status.

Among single-vehicle crashes in which the driver was fatigued in the pre-crash phase, 83.9 percent were ROR crashes, while among single-vehicle crashes in which the driver was not fatigued, only 55.5 percent were ROR crashes. The difference between the two percentages is statistically significant at the 90 percent confidence level (χ^2=9.39, p-value=0.0098). Therefore, it is more likely for a ROR crash to occur if the driver is fatigued as compared to if she/he is not.

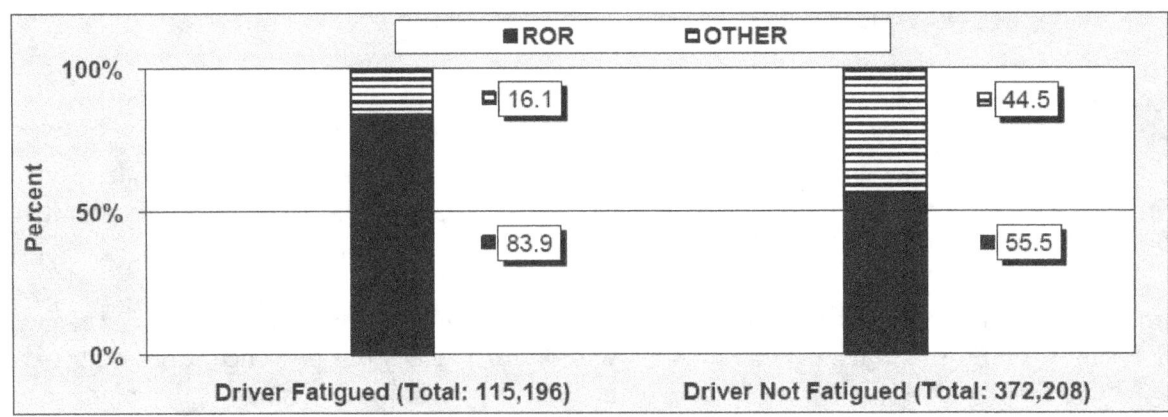

Figure 3: Single-Vehicle Crashes by Driver Fatigue Status

(Crashes With Missing and Unknown Values for the Factor Not Included)

Pre-Existing Physical or Mental Health Conditions

Figure 4 shows the frequencies of single-vehicle ROR and "Other" crashes by driver's pre-existing physical or mental health conditions. Among single-vehicle crashes in which the driver had pre-existing physical or mental health conditions, 75.6 percent were ROR crashes. This is significantly higher than the percentage (58.7%) of ROR crashes in which drivers did not have such conditions (χ^2=25.3, p-value=0.0003). This shows that drivers with pre-existing physical or mental health conditions are more likely to be involved in ROR crashes as compared to those who do not have such health conditions.

14

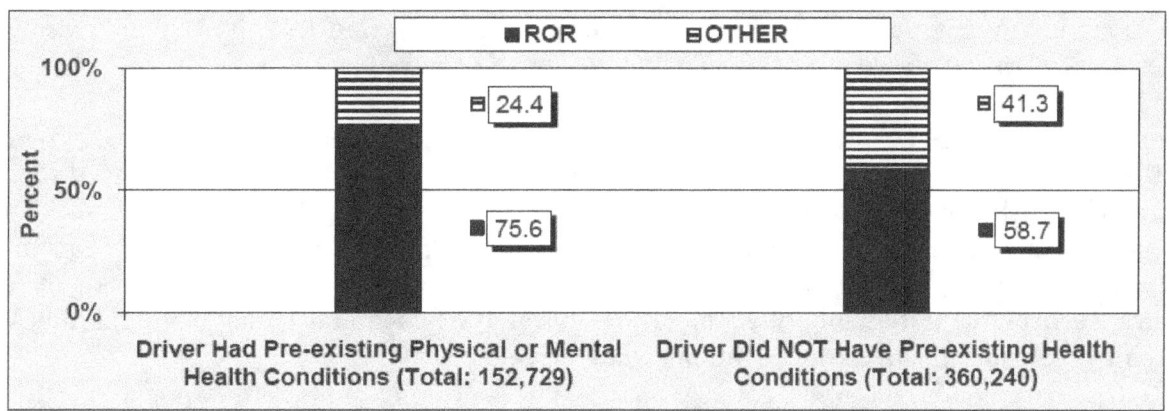

Figure 4: Single-Vehicle Crashes by Driver's Pre-Existing Physical or Mental Health Conditions

(Crashes With Missing and Unknown Values for the Factor Not Included)

Gender

Figure 5 shows that among single-vehicle crashes with male drivers, 68.1 percent were ROR crashes, while among single-vehicle crashes with female drivers, 61.1 percent were ROR crashes. The difference between these two percentages is statistically significant at the 90 percent confidence level (χ^2=5.17, p-value=0.0421). Thus, one could infer that the vehicles with male drivers are more likely to be involved in ROR crashes as compared to the vehicles with female drivers.

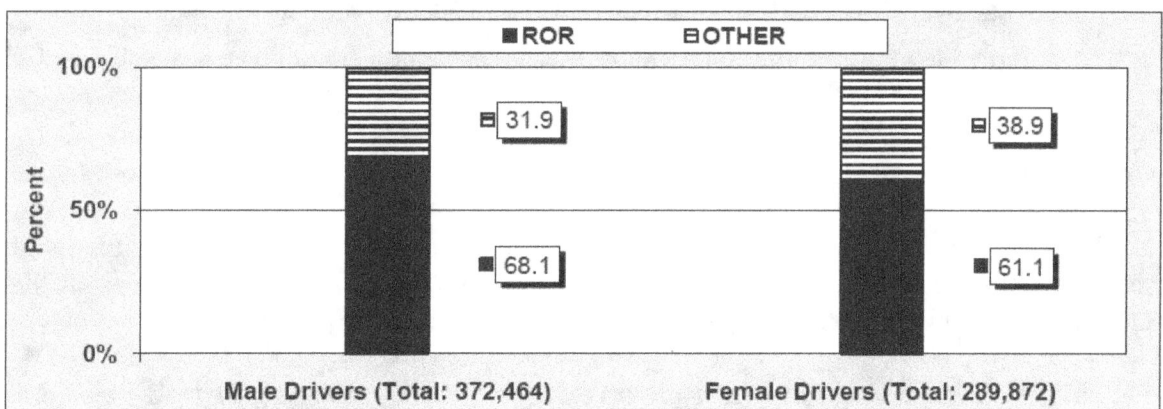

Figure 5: Single-Vehicle Crashes by Driver's Gender

(Crashes With Missing and Unknown Values for the Factor Not Included)

Familiarity With Roadway

In the NMVCCS, the driver's self-reported frequency of driving the roadway is used to define the driver's familiarity with the roadway on which the crash occurred. If the frequency of driving was daily, weekly, several times a month, or monthly, the driver is defined to be familiar with the roadway. Driving rarely or for the first time on the road indicates the driver's unfamiliarity

15

with the roadway. Figure 6 shows the frequencies of single-vehicle ROR and "Other" crashes by driver's familiarity with the roadway on which the crash occurred.

Among single-vehicle crashes in which the driver was familiar with the roadway, 63.9 percent were ROR crashes. When the driver was not familiar with the roadway, 54.1 percent of the single-vehicle crashes were ROR crashes. The difference between these two percentages is statistically significant at the 90 percent confidence level (χ^2=15.5, p-value=0.0020). The conclusion from the chi-square test, driving on familiar roadways is more likely to be involved in ROR crashes, seems to be not so intuitive. One of the possible explanations is that being familiar with the roadway makes the driver less cautious while driving.

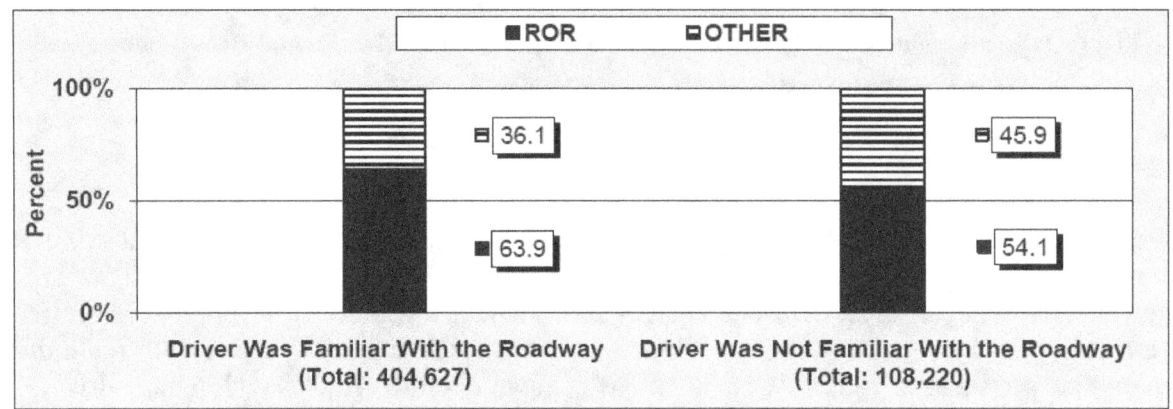

Figure 6: Single-Vehicle Crashes by Driver's Familiarity with the Roadway

(Crashes With Missing and Unknown Values for the Factor Not Included)

Work-Related Stress or Pressure

In the NMVCCS, driver's work-related stress or pressure was documented if the driver had been in this state in the days leading up to the crash. Figure 7 presents the frequencies of single-vehicle ROR and "Other" crashes by driver's work-related stress or pressure.

Among single-vehicle crashes in which the driver was feeling some type of work-related stress or pressure in the pre-crash phase, 86.4 percent were ROR crashes. However, only 59.5 percent were ROR crashes among single-vehicle crashes in which the driver was not feeling work-related stress or pressure. The difference between these two percentages is statistically significant at the 90 percent confidence level (χ^2=3.23, p-value=0.0973). Thus, one could infer that drivers experiencing work-related stress or pressure are more likely to be involved in ROR crashes.

16

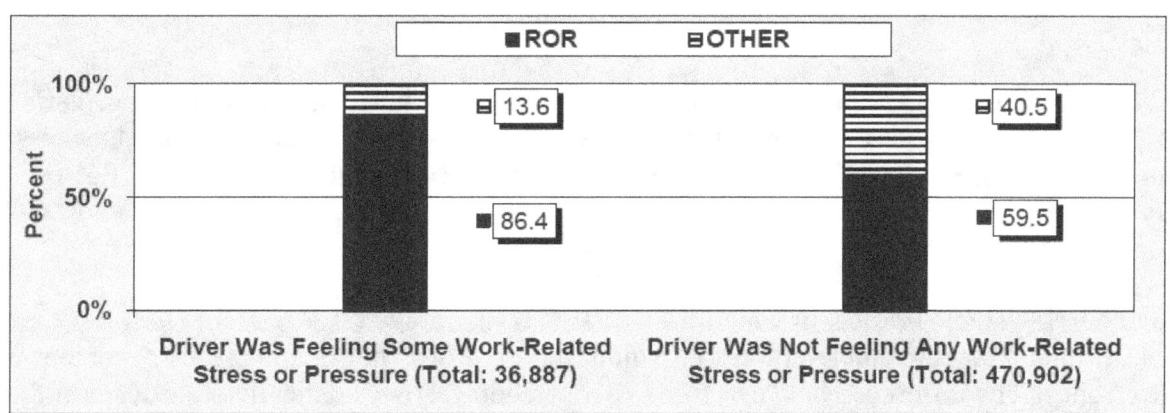

Figure 7: Single-Vehicle Crashes by Driver's Work-Related Stress or Pressure

(Crashes With Missing and Unknown Values for the Factor Not Included)

Roadway Surface Conditions

Two categories of roadway surface conditions are considered in this study: dry and wet (with standing water, snow, slush, or ice). The frequencies of single-vehicle ROR and "Other" crashes by roadway surface conditions are shown in Figure 8.

Among single-vehicle crashes in which the roadway surface was dry, 70.6 percent were ROR crashes. In contrast, among single-vehicle crashes in which the roadway surface was wet with water or ice or snow, only 47.3 percent were ROR crashes. The difference between these two percentages is statistically significant at the 90 percent confidence level (χ^2=58.19, p-value<0.0001). Thus, the occurrence of ROR crashes on wet roadways is less likely as compared to dry roadways. One possible explanation for this inference is that while driving on wet roadways drivers usually exercise more cautions.

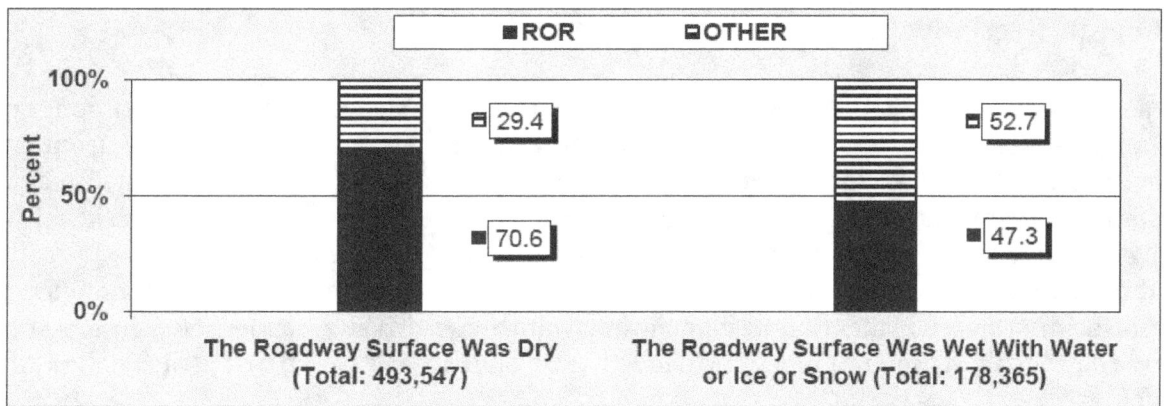

Figure 8: Single-Vehicle Crashes by Roadway Surface Conditions

(Crashes With Missing and Unknown Values for the Factor Not Included)

17

In a Hurry

The NMVCCS researchers identify whether a driver was in a hurry and provide the driver's reasons such as: late for start of work shift, late for start of school classes, late for business appointment, work related delivery schedule, late for social appointment, pursuing/fleeing, and normal driving pattern. Figure 9 presents the frequencies of single-vehicle ROR and "Other" crashes based on whether the driver was in a hurry or not.

Among single-vehicle crashes in which the driver was in a hurry, 82.9 percent were ROR crashes, while among single-vehicle crashes in which the driver was not in a hurry, 59.9 percent were ROR crashes. The difference between these two percentages is statistically significant at the 90 percent confidence level (χ^2=16.69, p-value=0.0015). Thus, drivers, when in a hurry, are more likely to be involved in ROR crashes.

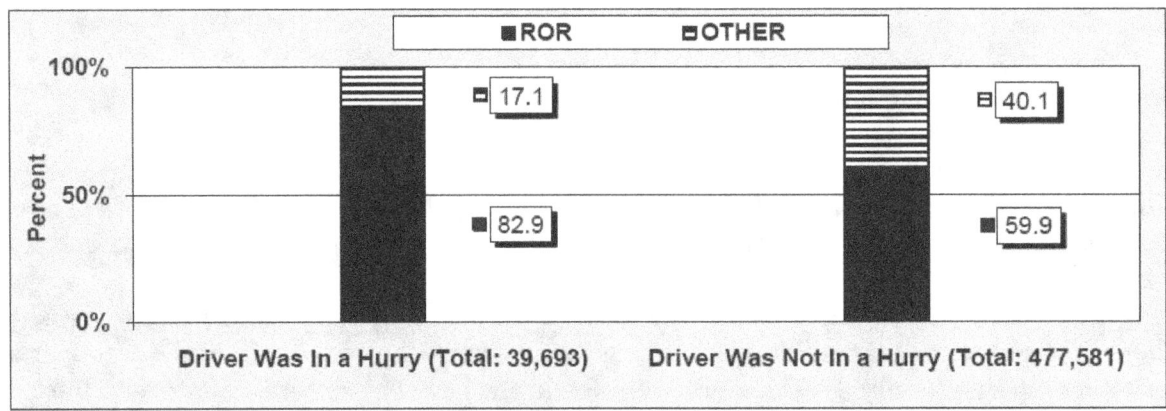

Figure 9: Single-Vehicle Crashes Based on Whether the Driver Was in a Hurry

(Crashes With Missing and Unknown Values for the Factor Not Included)

4.2 Logistic Regression Analysis

The analysis in Section 4.1 shows that crash-associated factors such as driver inattention, driver alcohol presence, driver fatigue status, and driver's gender, etc., are significant contributors to the occurrence of single-vehicle ROR crashes. It remains to assess their relative influence as well as estimate the amount of risk each carried (odds ratio) in the occurrence of such crashes. Logistic regression is used for this purpose. This is a statistical procedure that predicts the probability (p) of occurrence of an event (single-vehicle ROR crash, in the present case) as a consequence of certain factors (driver inattention, driver fatigue status, and driver's gender, etc., in the present case). The logistic regression model provides log of odds [= log (p/ (1-p))] as a function of the predictors:

Log (odds of single-vehicle ROR crash)
= a_0 + $a_1 \times$driver inattention + $a_2 \times$ driver alcohol presence + …+ $a_9 \times$driver's gender

where a_0 is the intercept and {$a_1, a_2, a_3 … a_9$} are the regression coefficients.

18

The odds ratio estimated in the logistic regression can tell a great deal about the risk a certain factor carried in contributing to the occurrence of ROR crashes. Odds ratio measures the magnitude of increase in odds of occurrence of an event as a result of a unit increase (for dichotomous variables, from 0 to 1) in a predictor.

The estimates of logistic regression coefficients and the corresponding odds ratio from the SAS SURVEYLOGISTIC procedure are shown in Table 8.

The ordered (decreasing) estimates of regression coefficients show that the most influential factor in the occurrence of single-vehicle ROR crashes was the factor "driver inattention," followed by "driver was fatigued," "driver was in a hurry," "the roadway surface was dry," "driver alcohol present," "driver was familiar with the roadway," "driver had pre-existing physical or mental health conditions," "driver was male," and "driver was feeling work-related stress or pressure."

Table 8 also shows odds ratio estimates of these crash-associated factors from the SAS logistic regression procedure.

- The odds ratio 3.66 for the factor "driver inattention" shows that the odds of being involved in an ROR crash for an inattentive driver were 3.66 times greater than the odds for an attentive driver.

- The odds ratio 3.48 for the factor "driver was fatigued" shows that the odds of being involved in an ROR crash for a fatigued driver were 3.48 times greater than the odds for a not fatigued driver.

- The odds ratio 3.20 for the factor "driver was in a hurry" shows that the odds of being involved in an ROR crash when driver was in a hurry were 3.20 times greater than the odds when driver was not in a hurry.

Similarly, the factors "the roadway surface was dry," "driver alcohol present," "driver was familiar with the roadway," "driver had pre-existing physical or mental health conditions," and "driver was male," were also statistically significantly linked with increased risk of single-vehicle ROR crash. The factor "driver was feeling work-related stress or pressure" was linked with the increased risk of single-vehicle ROR crash, though the link is not statistically significant at the 90 percent confidence level.

NHTSA's National Center for Statistics and Analysis   1200 New Jersey Avenue SE., Washington, DC 20590

Table 8: Logistic Regression Coefficients and Odds Ratios

Variable	Coefficient	Odds Ratio	*p*-value
Driver Inattention	1.2967	3.66	<.0001
Driver Was Fatigued	1.2463	3.48	<.0001
Driver Was In a Hurry	1.1630	3.20	<.0001
The Roadway Surface Was Dry	0.9928	2.70	<.0001
Driver Alcohol Present	0.9215	2.51	0.0218
Driver Was Familiar with the Roadway	0.7265	2.07	0.0032
Driver Had Pre-Existing Physical/Mental Health Conditions	0.5924	1.81	0.0031
Driver Was Male	0.2787	1.32	0.0217
Driver Was Feeling Work-Related Stress or Pressure	0.2252	1.25	0.5457

Data source: NMVCCS (2005 – 2007)

NHTSA's National Center for Statistics and Analysis   1200 New Jersey Avenue SE., Washington, DC 20590

5. The Effect of ABS and ESC on Run-Off-Road Crashes

Crash avoidance technologies are becoming increasingly useful in reducing traffic fatalities and injuries. In recent years, many new crash avoidance technologies such as lane departure systems and ESC systems have either been in the stages of design, development and refinement, or been already widely applied to the newer model vehicles. One primary objective of the NMVCCS is to help the highway safety community to evaluate and develop the vehicle-related crash avoidance technologies by identifying pre-crash events and factors leading up to a crash. This section investigates the role of two important technologies, ESC and ABS, in ROR crashes based on the NMVCCS data.

ABS is a four-wheel system that prevents wheel lock-up by automatically modulating the brake pressure during an emergency stop. By preventing the wheels from locking, it enables the driver to maintain steering control and stop in the shortest possible distance under most conditions. ESC is an evolution of the ABS concept. ESC uses a computerized technology that improves the safety of a vehicle's stability by detecting and preventing loss of control. According to FMVSS 126, all new passenger vehicles are required to be equipped with ESC after September 1, 2011.

Previous studies in the early 1990s showed significant increases in fatal run-off-road crashes involved ABS-equipped vehicles.[11] A second generation of analyses around the year of 2000 showed much smaller increases in run-off-road crashes with ABS and the increases were no longer statistically significant. A recent NHTSA study[12] evaluating the long-term effect of ABS in passenger vehicles reported that ABS had a close to zero net effect on fatal crashes (FARS, 1995-2007) but ABS was quite effective in non-fatal crashes (National Automotive Sampling System General Estimates System (NASS-GES), 1995-2007) at the 90 percent confidence level, reducing the overall crash involvement rate by 6 percent in passenger cars and by 8 percent in light trucks. The reduced significance of ABS's involvement in ROR crashes along the timeline of the studies in the past two decades may be due to the following two reasons: (1) drivers have known better how ABS works; and (2) more important, the sample sizes for the later studies were much larger. This most recent NHTSA report also studied the combined effect of ESC and ABS and concluded that ESC along with ABS would prevent a large proportion of fatal and non-fatal crashes.[12]

The EQUIP (Equipment) file in the NMVCCS dataset has collected information on the availability and use of equipments on-board the vehicles, both original equipment (OEM) and aftermarket. There is one record for each type of pre-listed equipment (which may or may not be present on-board) for each case vehicle. This makes it possible to evaluate the effect of ABS and/or ESC on fatal and non-fatal run-off-road crashes. In this section, the estimated percentages of ROR crash involvement of passenger vehicles equipped with ABS and/or ESC* are compared with those of the passenger vehicles not equipped with the technologies.

---

* In the NMVCCS, the availability and use of ESC reflects mostly standard ESC because it is difficult to know whether a particular vehicle has optional ESC installed.

NHTSA's National Center for Statistics and Analysis   1200 New Jersey Avenue SE., Washington, DC 20590

To overcome the small sample size limitations in estimating, the evaluation of ABS and ESC in this section includes all passenger vehicles in both single- and multiple-vehicle ROR crashes. Discussions of ROR crashes in previous sections of this report concern single-vehicle crashes only unless otherwise stated.

ABS Versus "Neither ABS Nor ESC"

The percentages of vehicles equipped with ABS versus "neither ABS nor ESC" involved in ROR crashes are displayed in Figure 10. It shows that for the vehicles equipped with ABS, 13.2 percent ran off the road, while for the vehicles equipped with neither ABS nor ESC, 14.6 percent ran off the road. The odds ratio of 1.1[†] shows that the odds of being involved in ROR crashes for the vehicles equipped with neither ABS nor ESC were 1.1 times greater than the odds for the vehicles equipped with ABS. This positive effect of ABS on reducing the ROR crashes is not statistically significant at the 90 percent confidence level (p-value=0.500).

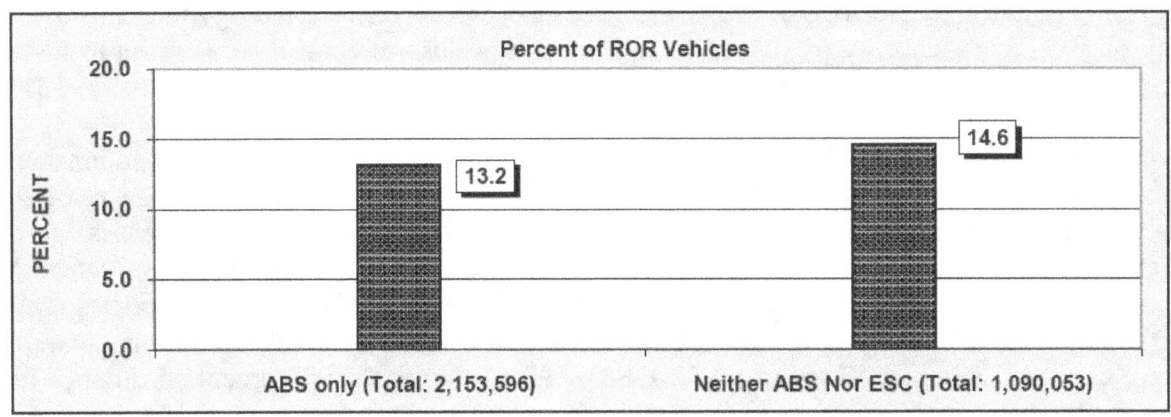

Figure 10: ROR Crash Involvement Among Vehicles Equipped With ABS Versus "Neither ABS Nor ESC"

"ABS Only" Versus "Both ABS and ESC"

The percentages of vehicles equipped with "ABS only" versus "both ABS and ESC" involved in ROR crashes are shown in Figure 11. It shows that for the vehicles equipped with both ABS and ESC, 7.5 percent ran off the road, while for the vehicles equipped with ABS only, 13.2 percent ran off the road.

The odds ratio of 1.9[‡] shows that the odds of being involved in ROR crashes for the vehicles equipped with ABS only were 1.9 times greater than the odds for the vehicles equipped with both ABS and ESC. There was a positive effect of ESC on reducing the ROR crashes, though the association is not statistically significant at the 90 percent confidence level (p-value=0.154).

[†] $1.1 = [0.146/(1-0.146)]/[0.132/(1-0.132)]$
[‡] $1.9 = [0.132/(1-0.132)]/[0.075/(1-0.075)]$

NHTSA's National Center for Statistics and Analysis   1200 New Jersey Avenue SE., Washington, DC 20590

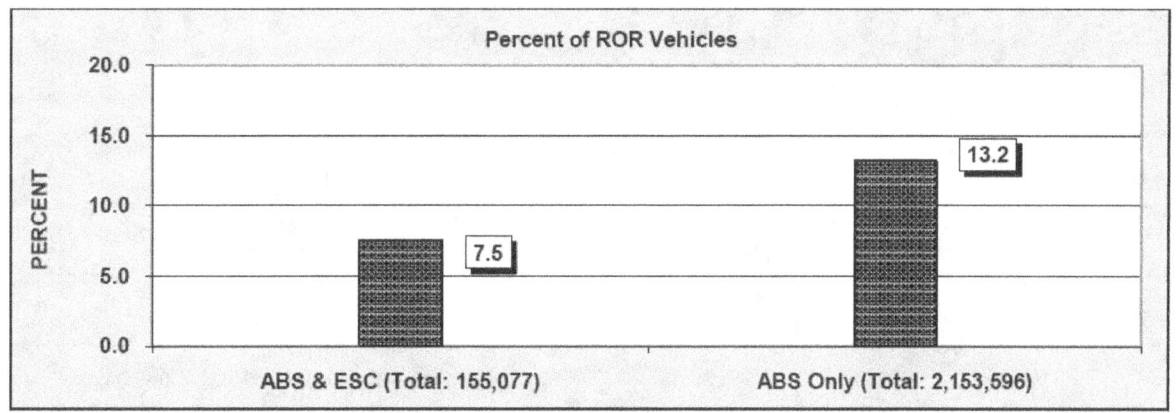

Figure 11: ROR Crash Involvement Among Vehicles Equipped With "ABS Only" Versus "Both ABS And ESC"

"Both ABS and ESC" Versus "Neither ABS Nor ESC"

The percentages of vehicles equipped with "both ABS and ESC" versus "neither ABS nor ESC" involved in ROR crashes are presented in Figure 12. It shows that for the vehicles equipped with both ABS and ESC, 7.5 percent ran off the road, while for the vehicles equipped with neither ABS nor ESC, 14.6 percent ran off the road.

The odds ratio of 2.1 § shows that the odds of being involved in ROR crashes for the vehicles equipped with neither ABS nor ESC were 2.1 times greater than the odds for the vehicles equipped with both ABS and ESC. This is statistically significant at the 90 percent confidence level (p-value=0.065). Thus, one could infer that the combined effect of ABS and ESC systems on reducing the ROR crashes is significant. This result agrees with the finding in the aforementioned NHTSA report[12], which, based on the data from FARS and GES, claimed that the combination of ABS and ESC would reduce fatal ROR crashes by an estimated 30 percent in passenger cars and by 68 percent in light trucks.

---

§ $2.1 = [0.146/(1-0.146)]/[0.075/(1-0.075)]$

23

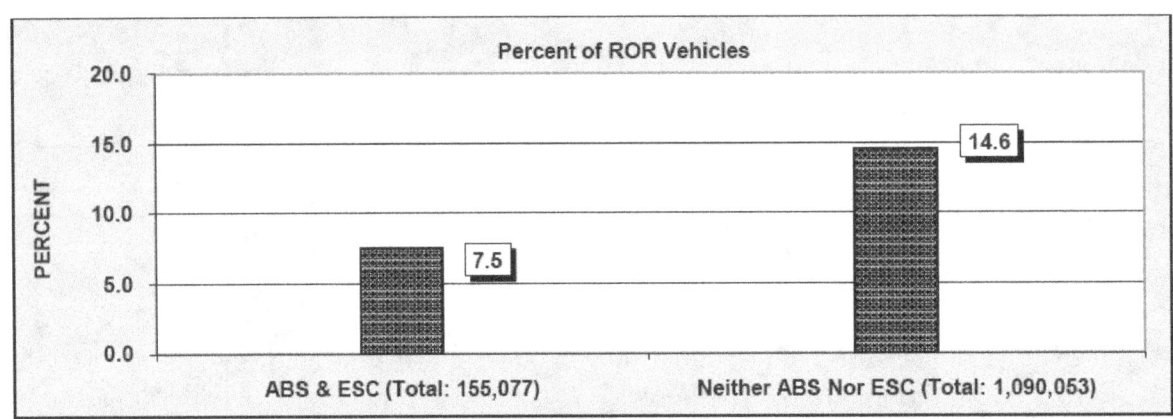

Figure 12: ROR Crash Involvement Among Vehicles Equipped With "Both ABS and ESC" versus "Neither ABS Nor ESC"

Discussions

One caveat of this study concerning ABS and ESC is that the ROR vehicles are from the NMVCCS data collected during a period of only 2.5 years from July 3, 2005, to December 31, 2007. However, the findings from this study are consistent with the previously published results on the evaluation of the long-term effects of ABS and ESC in passenger vehicles based on the 13 years of FARS and GES data, especially on the combined effect of ABS and ESC. It has confirmed the usefulness of the NMVCCS data in evaluating crash avoidance technologies. In fact, this study is NHTSA's very first effort in evaluating the effectiveness of crash avoidance technologies with the NMVCCS data.

24

6. Summary and Conclusions

Run-off-road crashes account for a significant percentage of single-vehicle crashes. This study has thoroughly investigated vehicle-, weather-, roadway-, and driver-related factors in single-vehicle ROR crashes with a focus on driver's physical and mental conditions as well as driver's activities prior to the crash. This was done by using the NMVCCS data that provides on-scene information about crashes.

Among all critical reasons for passenger vehicle single-vehicle ROR crashes, more than 95 percent were driver-related. The dominant critical reasons (in descending order) were "internal distraction," "overcompensation," "poor directional control," "too fast for curve," and "sleeping/actually asleep." In comparison, for large trucks (based on the LTCCS), "sleeping/actually asleep" and "heart attack or other physical impairment" were the most frequently assigned critical reasons for the single-vehicle ROR crashes. Therefore, although ROR crash countermeasures (like improving roadways as recommended by previous studies) might work to some extent, ROR crash prevention efforts should focus more on drivers. The public should be more aware of the dangers of irresponsible driving behaviors such as distracted driving, fatigued driving, and speeding.

The logistic regression analysis shows that the most influential factors in the occurrence of single-vehicle ROR crashes are the factors "driver inattention," "driver was fatigued," and "driver was in a hurry."

As a caution, it should be noted that the terms "critical event," "critical reason," and "associated factors" used in the NMVCCS are not indicative of the cause of a crash. Although the NMVCCS collected information on many factors that may have contributed to a crash, none of them should be considered as the single ultimate cause of the crash. In addition, since the NMVCCS data is based on some sample design, all estimates are subject to sampling error.

This study has also tried to evaluate the role of ABS and ESC in reducing ROR crashes. It shows that the odds of being involved in ROR crashes for the vehicles equipped with neither ABS nor ESC were 2.1 times greater than the odds for the vehicles equipped with both ABS and ESC. The combined effect of ABS and ESC systems on reducing the ROR crashes is significant, which is consistent with NHTSA's evaluation of long-term effect of ABS and ESC in passenger vehicles based on the FARS and GES data. This last evaluation demonstrated the usefulness of the NMVCCS data in evaluating crash avoidance technologies. This study is NHTSA's first effort in evaluating the effectiveness of crash avoidance technologies based on the NMVCCS data.

NHTSA's National Center for Statistics and Analysis   1200 New Jersey Avenue SE., Washington, DC 20590

7. References

[1]   NHTSA. (2009). *Traffic Safety Facts 2008: A Compilation of Motor Vehicle Crash Data from the Fatality Analysis Reporting System and the General Estimates System.* DOT HS 811 170. Washington, DC: National Highway Traffic Safety Administration. Available at http://www-nrd.nhtsa.dot.gov/pubs/811170.pdf

[2]   Spainhour, L. K., & Mishra, A. (2007). *Analysis of Fatal Run-off-the-Road Crashes Involving Overcorrection.* TRB 87th Annual Meeting, Paper No. 08-0661. Washington, DC: Transportation Research Board.

[3]   Örnek, E.. & Drakopoulos, A. (2007, August). *Analysis of Run-Off-Road Crashes in Relation to Roadway Features and Driver Behavior.* Proceedings of the 2007 Mid-Continent Transportation Research Symposium, Ames, Iowa.

[4]   Johnston, I., Corben, B., Triggs, T., Candappa, N. & Lenné, M. (2006, June). *Reducing Serious Injury and Death from Run-off-Road Crashes in Victoria – Turning Knowledge into Action.* RACV Research Report. Victoria, Australia: Royal Automobile Club of Victoria Ltd.

[5]   Liu, C., & Subramanian, R. (2009, November). *Factors Related to Fatal Single-Vehicle Run-Off-Road Crashes.* DOT HS 811 232. Washington, DC: National Highway Traffic Safety Administration. Available at http://www-nrd.nhtsa.dot.gov/Pubs/811232.pdf

[6]   NHTSA. (2008, July). *National Motor Vehicle Crash Causation Survey – Report to Congress.* DOT HS 811 059. Washington, DC: National Highway Traffic Safety Administration. Available at http://www-nrd.nhtsa.dot.gov/Pubs/811059.PDF

[7]   FMCSA . (2006, March). *Report to Congress on the Large Truck Crash Causation Study.* Washington, DC: Federal Motor Carrier Safety Administration. March 2006. Available at http://www.fmcsa.dot.gov/facts-research/research-technology/report/ltccs-2006.htm#EXECSUM

[8]   Craft, R. H., & Preslopsky, B. (2009, September). *Driver Distraction and Inattention in the USA Large-Truck and National Motor Vehicle Crash Causation Studies.* First International Conference on Driver Distraction and Inattention, Gothenburg, Sweden.

[9]   Starnes, M. (2006, August). *Large-Truck Crash Causation Study – An Initial Overview.* DOT HS 810 646. Washington, DC: National Highway Traffic Safety Administration. Available at http://www-nrd.nhtsa.dot.gov/Pubs/810646.PDF

[10]   McCartt, A. T., Rohrbaugh, J. W., Hammer, M. C. & Fuller, S. Z. (2000). *Factors Associated with Falling Asleep at the Wheel Among Long-Distance Truck Drivers*, Accident Analysis and Prevention, 32, 493-504.

[11]   Hertz, E., Hilton, J. and Johnson, D.M. (1998, August). *Analysis of the Crash Experiences of Vehicles Equipped with Antilock Braking Systems: An Update.* DOT HS 808 758. Washington, DC: National Highway Traffic Safety Administration. Available at http://www-nrd.nhtsa.dot.gov/Pubs/808758.PDF

[12] Kahane, C. J., & Dang, J. N. (2009, August). *The Long-Term Effect of ABS in Passenger Cars and LTVs*. DOT HS 811 182. Washington, DC: National Highway Traffic Safety Administration. Available at http://www-nrd.nhtsa.dot.gov/Pubs/811182.pdf

27

DOT HS 811 500
July 2011

U.S. Department
of Transportation

**National Highway
Traffic Safety
Administration**

www.nhtsa.gov